Museum Indian

Zoology of the Royal Indian Marine Survey ship Investigator

Under the command of Commander T.H. Hemming

Museum Indian

Zoology of the Royal Indian Marine Survey ship Investigator
Under the command of Commander T.H. Hemming

ISBN/EAN: 9783337269128

Printed in Europe, USA, Canada, Australia, Japan

Cover: Foto ©berggeist007 / pixelio.de

More available books at **www.hansebooks.com**

ILLUSTRATIONS

OF THE

ZOOLOGY

OF THE

ROYAL INDIAN MARINE SURVEY SHIP

INVESTIGATOR,

UNDER THE COMMAND OF

COMMANDER T. H. HEMING R.N.

FISHES.—Part V, Plates XVIII—XXIV.
CRUSTACEA.—Part VI, Plates XXXIII—XXXV.
MOLLUSCA.—Part II, Plates VII—VIII.

UNDER THE DIRECTION OF

A. ALCOCK, M. B., C.M.Z.S., SUPERINTENDENT OF THE INDIAN MUSEUM, FORMERLY
SURGEON-NATURALIST TO THE INDIAN MARINE SURVEY

AND OF

A. R. S. ANDERSON, B.A., M.B., SURGEON-NATURALIST TO THE INDIAN MARINE SURVEY

CALCUTTA:
OFFICE OF THE SUPERINTENDENT OF GOVERNMENT PRINTING, INDIA.
1898.

EXPLANATION OF PLATE.

FISHES.

PLATE XVIII.

Fig. 1.—*Bathyscriola cyanea*, Alcock, Annals and Magazine of Natural History, (6), VI, 1890, p. 202, and (6), VIII, 1891. p. 23.

Fig. 2.—*Brephostoma carpenteri*, Alcock, Annals and Magazine of Natural History, (6), IV, 1889, p. 383, and (6), VI, 1890, p. 201.

Fig. 3.—*Minous inermis*, Alcock, Journal, Asiatic Society of Bengal, Vol. LVIII, pt. 2, 1889, p. 299. Annals and Magazine of Natural History, (6), X, 1892, p. 207. Journal, Asiatic Society of Bengal, Vol. LXIII, pt. 2, 1894, p. 116.

Fig. 4.—*Pterois macrura*, Alcock, Journal, Asiatic Society of Bengal, Vol. LXV, pt. 2, 1896, p. 303.

Fig. 5.—*Sebastes unciparus*, Alcock, Journal, Asiatic Society of Bengal, Vol. LVIII, pt 2, 1889, p. 298.

Fig. 6.—*Scorpæna crostris*, Alcock, Journal, Asiatic Society of Bengal, Vol. LXV, pt. 2, 1896, p. 302.

Fig. 7.—*Scorpæna bucephalus*, Alcock, Journal, Asiatic Society of Bengal, Vol. LXV, pt. 2, 1896, p. 302.

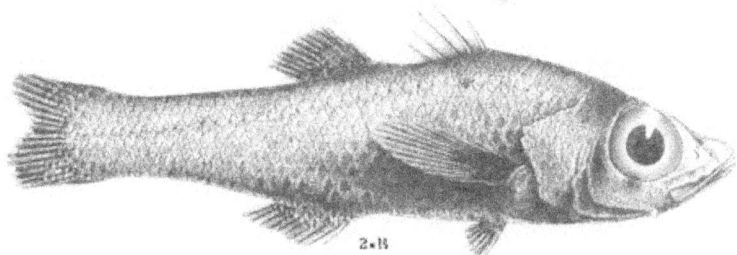

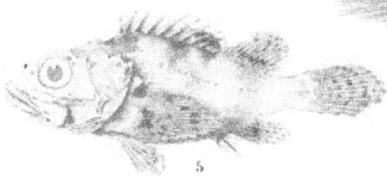

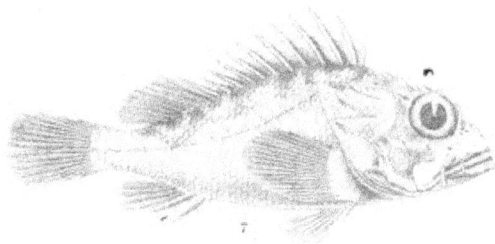

1 Bathysphon evanes 2 Bembatoma carpenteri 3 Minous inermis 4 Pterois macrura

EXPLANATION OF PLATE.

FISHES.

PLATE XIX.

Fig. 1.—*Halieutæa coccinea*, Alcock, Annals and Magazine of Natural History, (6), IV, 1889, p. 382. Journal, Asiatic Society of Bengal, Vol. LXIII, pt. 2, 1894, p. 120.

Fig. 2.—*Halieutæa nigra*, Alcock, Annals and Magazine of Natural History, (6), VIII, 1891, p. 24. Journal, Asiatic Society of Bengal, Vol. LXIII, pt. 2, 1894, p. 120.

Fig. 3.—*Lophius indicus*, Alcock, Journal, Asiatic Society of Bengal, Vol. LVIII, pt. 2, 1889, p. 302.

Fig. 4.—*Malthopsis lutea*, Alcock, Annals and Magazine of Natural History, (6), VIII, 1891, p. 26.

Fig. 5.—*Haliemetus ruber*, Alcock, Annals and Magazine of Natural History (6), VIII, 1891, p. 27.

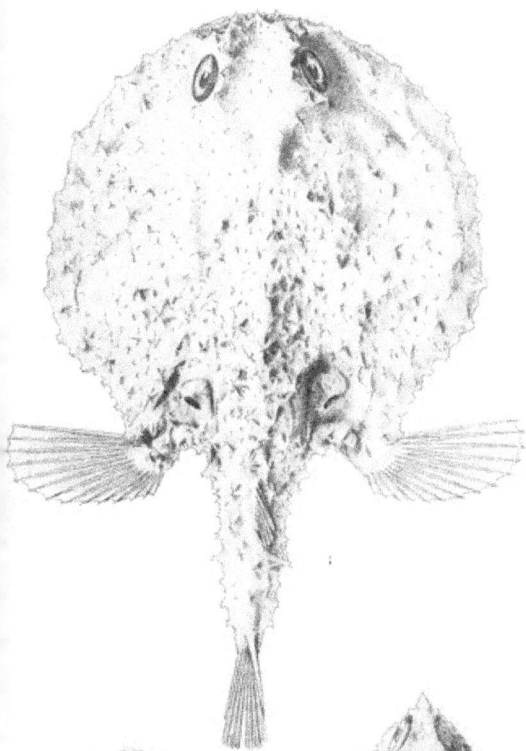

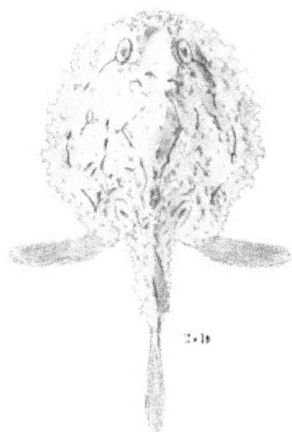

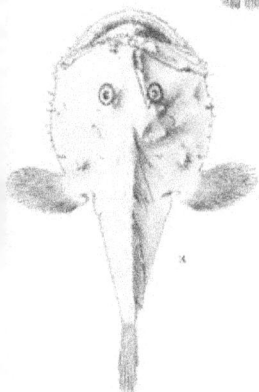

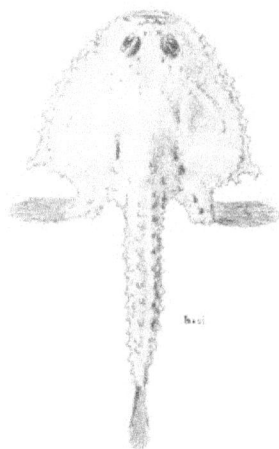

EXPLANATION OF PLATE.

FISHES.

PLATE XX.

Fig. 1.—Dibranchus micropus, Alcock, Annals and Magazine of Natural History, (6) VIII, 1891, p. 25 ; and (6) X, 1892, p. 348.

Fig. 2.—Dibranchus nasutus, Alcock, Annals and Magazine of Natural History, (6) VIII, 1891, p. 24.

Fig. 3.—Gobius cometes, Alcock, Annals and Magazine of Natural History, (6) VI, 1890, p. 208.

Fig 4.—Callionymus carebares, Alcock, Annals and Magazine of Natural History, (6) VI, 1890, p. 209.

Fig. 5.—Tæniolabrus cyclograptus, Alcock, Annals and Magazine of Natural History, (6) VI, 1890, p. 430.

Fig. 6.*—Bembrops gobioides (Goode), Proc. U. S. Nat. Mus., III, 1880, p. 347 : Goode and Bean, Oceanic Ichthyology. p. 290. *Vide* Alcock, Annals and Magazine of Natural History, (7) II, pp. 141, 142.

Fig. 7.—Amblyopus arctocephalus, Alcock, Annals and Magazine of Natural History, (6) VI, 1890, p. 432.

---

* *Hypsicometes gobioides* Goode, = *Bembrops platyrhynchus* Alcock.

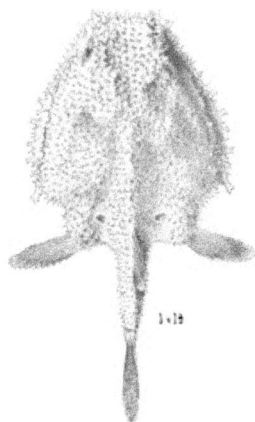

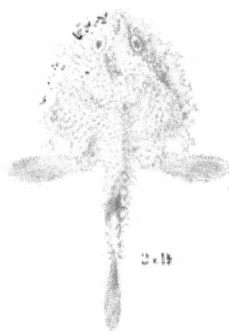

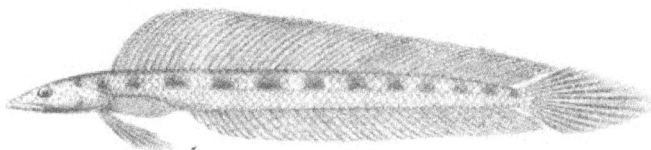

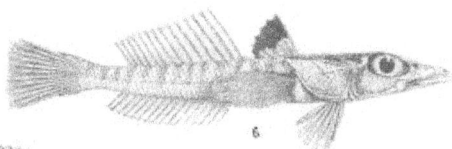

1 Dibranchus micropus   2 Dibranchus nasutus   3 Gobius cometes   4 Callionymus
5 Taeniolabrus cyclograptus   6 Bembrops platyrhynchus   7 Amblyopus arctocephalus

EXPLANATION OF PLATE.

FISHES.

PLATE XXI.

Fig. 1.—*Neobythites squamipinnis*, Alcock, Annals and Magazine of Natural History, (6) IV, 1889, p. 386 : Journal, Asiatic Society of Bengal, Vol. LXIII, pt. 2, 1894, p. 123.

Fig. 2.—*Neobythites steatiticus*, Alcock, Journal, Asiatic Society of Bengal, Vol. LXII, pt. 2, 1893, p. 181.

Fig. 3.—*Tauredophidium hextii*, Alcock, Annals and Magazine of Natural History, (6) VI, 1890, p. 213.

Fig. 4.—*Dermatorus melanocephalus*, Alcock, Annals and Magazine of Natural History, (6) VIII, 1891, p. 32.

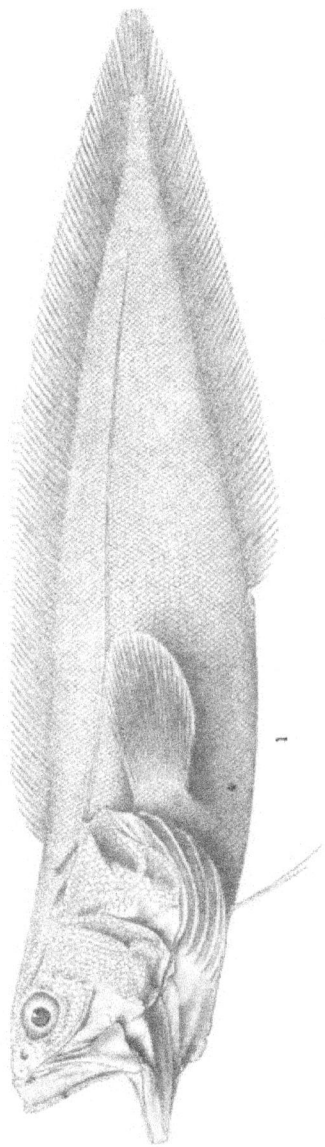

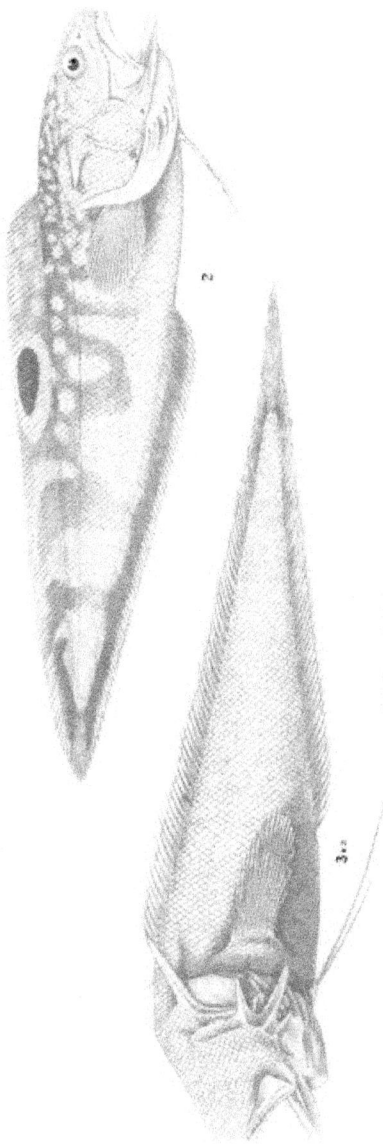

EXPLANATION OF PLATE.

FISHES.

PLATE XXII.

Fig. 1.—*Physiculus argyropastus*, Alcock, Journal, Asiatic Society of Bengal, Vol. LXII, pt. 2, 1893, p. 180, and Vol LXIII, pt. 2, 1894, p. 121.

Fig. 2.—*Brachypleura xanthosticta*, Alcock, Journal, Asiatic Society of Bengal, Vol. LVIII, pt. 2, 1889, p. 281.

Fig. 3.—*Hephthocara simum*, Alcock, Annals and Magazine of Natural History, (6) X, 1892, p. 349.

Fig. 4.—*Arnoglossus brevirictis*, Alcock, Annals and Magazine of Natural History, (6) VI, 1890, p. 433.

Fig. 5.—*Cynoglossus carpenteri*, Alcock, Journal, Asiatic Society of Bengal, Vol. LVIII, pt. 2, 1889, p. 287: Annals and Magazine of Natural History, (6) VI, 1890, p. 217.

EXPLANATION OF PLATE.

FISHES.

PLATE XXIII.

Fig. 1.—*Sciauectes macrophthalmus*, Alcock, Annals and Magazine of Natural History, (6) VI, 1890, p. 216.

Fig. 2.—*Samaris cristatus*, Gray, Zoological Miscellany, p. 4 ; Günther, Cat. Fishes, Brit. Mus., IV, 420, Alcock, Journal, Asiatic Society of Bengal, Vol. LVIII, pt. 2, 1889, p. 291.

Fig. 3.—*Arnoglossus macrolophus*, Alcock, Journal, Asiatic Society of Bengal, Vol. LVIII, pt. 2, 1889, p. 280 ; Annals and Magazine of Natural History, (6) VI, 1890, p. 433.

Fig. 4.——*Læops güntheri*, Alcock, Annals and Magazine of Natural History, (6) VI, 1890, p. 438.

Fig. 5.—*Cynoglossus versicolor*, Alcock, Annals and Magazine of Natural History, (6) VI, 1890, p. 442.

Fig. 6.—*Cynoglossus præcisus*, Alcock, Annals and Magazine of Natural History, (6) VI, 1890, p. 442.

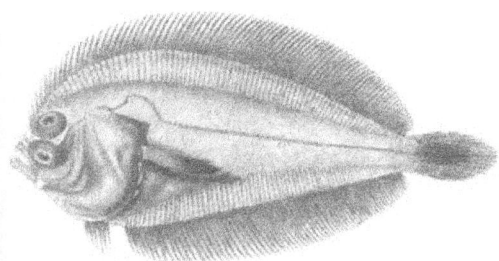

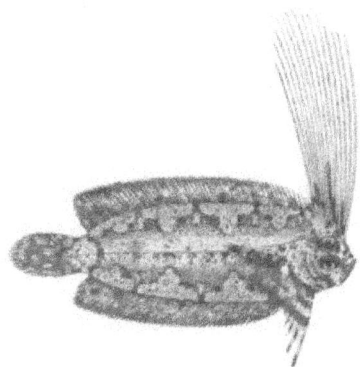

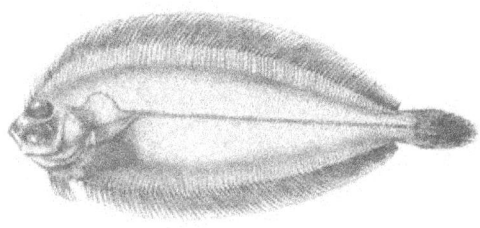

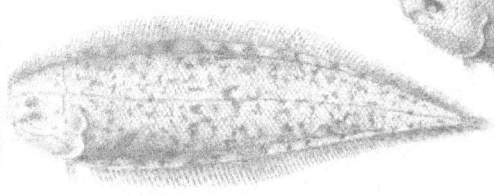

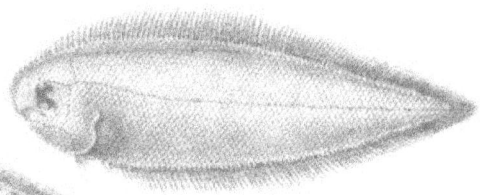

EXPLANATION OF PLATE.

FISHES.

PLATE XXIV.

Fig. 1.—*Synaptura altipinnis,* Alcock, Annals and Magazine of Natural History, (6) VI, 1890, p. 441.

Fig. 2.—*Rhomboidichthys valderostratus,* Alcock, Annals and Magazine of Natural History, (6) VI, 1890, p. 435.

Fig. 3.—*Rhomboidichthys azureus,* Alcock, Journal, Asiatic Society of Bengal, Vol. LVIII, pt. 2, 1889, p. 283: Annals and Magazine of Natural History, (6) VI, 1890, p. 435.

Figs. 4, 5.—*Rhomboidichthys polylepis,* Alcock, Annals and Magazine of Natural History, (6) VI, 1890, p. 434. (Fig. 4, young, fig. 5, adult female.)

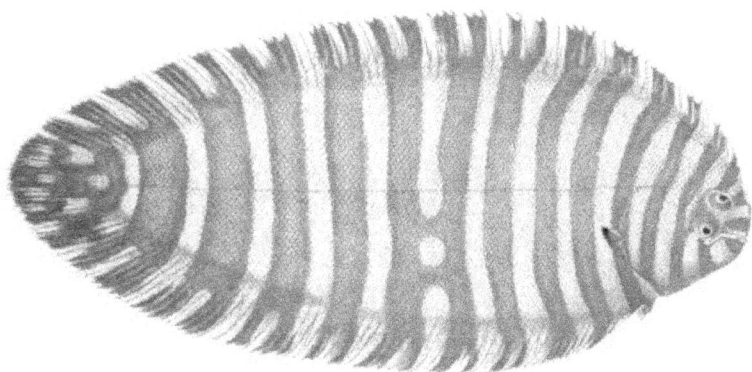

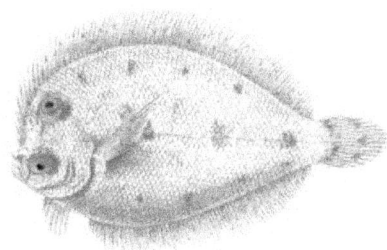

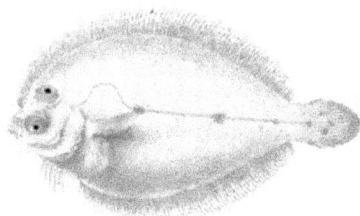

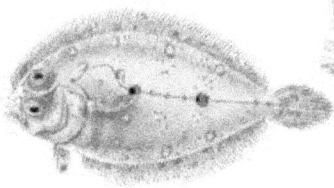

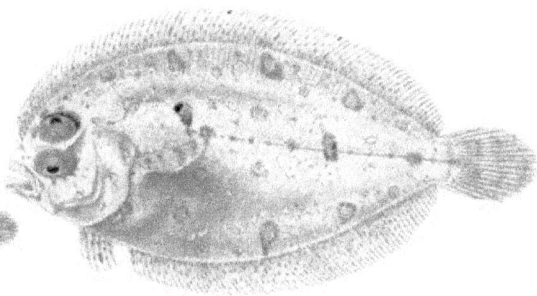

1 Synaptura dupanus 2 Rhomboidichthys valderostratus 3 Rhomboidichthys azurens
4 Anoglossus polylepis 5 Rhomboidichthys polylepis

EXPLANATION OF PLATE.

CRUSTACEA.

PLATE XXXIII.

Figs. 1, 1*a***.**—*Achaeus lacertosus*, Stimpson, Proc., Academy Natural Science, Philad. 1857, p. 218. [*Vide* Alcock, Journal, Asiatic Society of Bengal, Vol. LXIV, pt. 2, 1895, p. 172, *Carcinological Fauna of India*.]

Figs. 2, 2*a***.**—*Naxia cerastes*, Ortmann, in Semon's Forschungsr. Austral. u. Malay. Arch., Crust. (Jena. Denk. Bd., VIII) p. 43. [*Vide* Alcock, Carcinol. Faun. Ind. *loc. cit.*, p. 220.]

Figs. 3, 3*a***.**—*Xenocarcinus tuberculatus*, White, Proc. Zool. Soc. 1847, p. 119. [*Vide* Alcock, Carcinol. Faun. Ind. *loc. cit.*, p. 192.]

Figs. 4, 4*a***.**—*Hyastenus tenuicornis*, Pocock, Annals and Magazine of Natural History, (6) V, 1890, p. 76. [*Vide* Alcock, Carcinol. Faun. Ind. *loc. cit.*, p. 215.]

Figs. 5, 5*a***.**—*Naxia taurus*, Pocock, Annals and Magazine of Natural History, (6) V, 1890, p. 77. [*Vide* Alcock, Carcinol. Faun. Ind. *loc. cit.*, p. 219.]

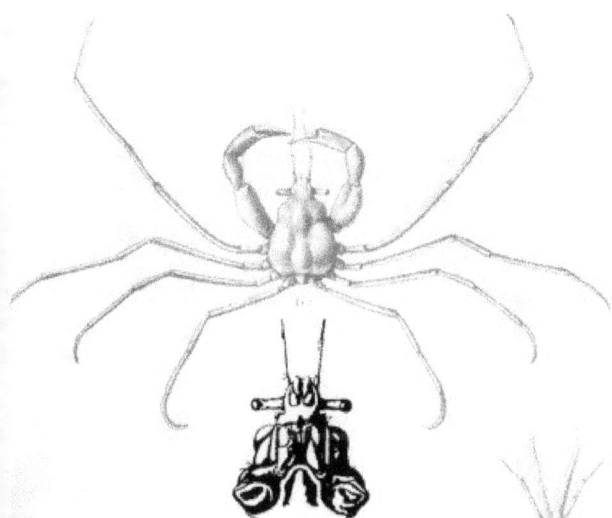

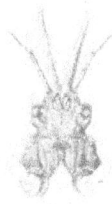

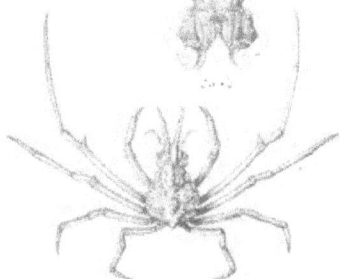

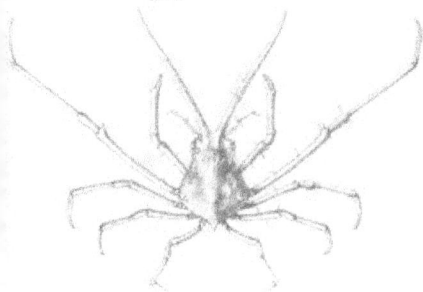

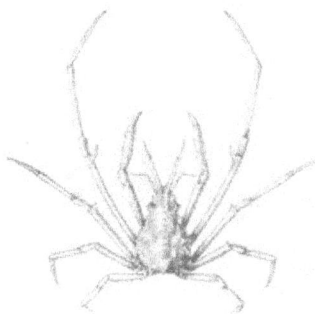

EXPLANATION OF PLATE.

CRUSTACEA.

PLATE XXXIV.

Figs. 1, 1a ♂, 2, 2a ♀.—Hoplophrys oatesii, Henderson, Trans. Linn. Soc., Zool., (2) V, 1893, p. 347. [*Vide* Alcock, Carcinol. Faun. Ind., *loc. cit.*, p. 233.]

Figs. 3, 3a,—Maia spinigera, DeHaan, Fauna Japonica Crust., p. 93. [*Vide* Alcock, Carcinol. Faun. Ind., *loc. cit.*, p. 239.]

Figs. 4, 4a, 4b.—Chlorinoides longispinus (DeHaan), Fauna Japonica, Crust., p. 94. [*Vide* Alcock, Carcinol. Faun. Ind., *loc. cit.*, p. 242.]

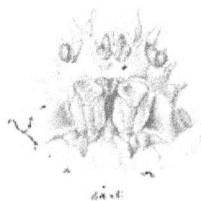

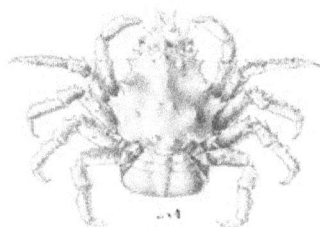

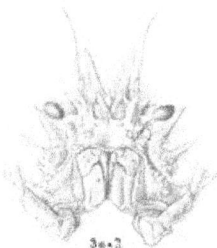

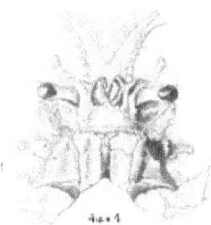

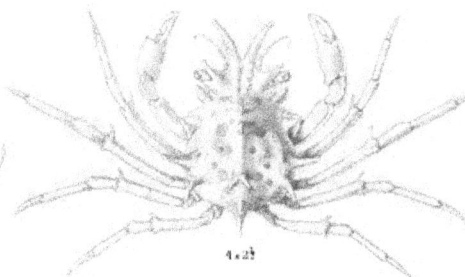

EXPLANATION OF PLATE.

CRUSTACEA.

PLATE XXXV.

Figs. 1, 1a.—*Schizophrys aspera*, Milne Edwards, Hist. Nat. Crust. I, 320. [*Vide* Alcock, Carcinol. Faun. Ind., *loc. cit.*, p. 243.]

Figs. 2, 2a.—*Schizophrys dama* (Herbst), Krabben und Krebse, III, iv, p. 5. [*Vide* Alcock, Carcinol. Faun. Ind., *loc. cit.*, p. 245.]

Figs. 3, 3a.—*Micippe margaritifera*, Henderson, Trans. Linn. Soc., Zool., (2) V, 1893, p. 348. [*Vide* Alcock, Carcinol. Faun. Ind., *loc. cit.*, p. 253.]

Figs. 4, 4a.—*Micippe margaritifera*, var. *parca*, Alcock, Carcinol. Faun. Ind., *loc. cit.*, p. 253.

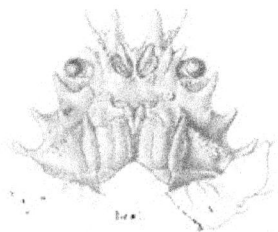

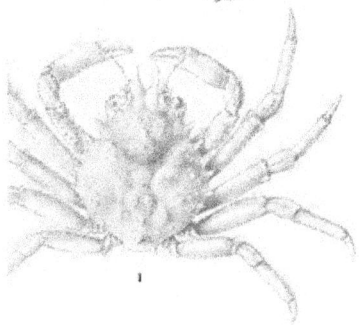

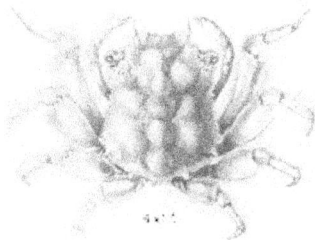

EXPLANATION OF PLATES.

MOLLUSCA.

PLATE VII.

Figs. 1, 1a.—*Pleurotoma (Ancistrosyrinx) travancorica*, E. A. Smith, Annals and Magazine of Natural History, (6) XVIII, 1896, p. 368.

Figs. 2, 2a.—*Pleurotoma (Surcula) profundorum*, E. A. Smith. Annals and Magazine of Natural History, (6) XVIII, 1896, p. 369.

Figs. 3, 3a.—*Natica (Lunatia) abyssicola*, E. A. Smith, Annals and Magazine of Natural History, (6) XVIII, 1896, p. 370.

Figs. 4, 4a.—*Natica (Lunatia) levis*, E. A. Smith, Annals and Magazine of Natural History, (6) XVIII, 1896, p. 370.

Figs. 5, 5a.—*Dentalium magnificum*, E. A. Smith, Annals and Magazine of Natural History, (6) XVIII, 1896, p. 371.

Figs. 6, 6a.—*Cardium (Fragum) simillimum*, E. A. Smith, Annals and Magazine of Natural History, (6) XVIII, 1896, p. 372.

 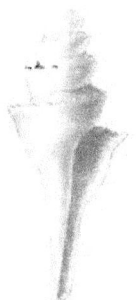  

1.2   1a.2   1.2   2a.2

3.3   3a.3   4.2   4a.2

5   6.3

EXPLANATION OF PLATES.

MOLLUSCA.

PLATE VIII.

Figs. 1, 1a, 1b, 1c.—*Yoldia anatina*, E. A. Smith, Annals and Magazine of Natural History, (6) XVIII, 1896, p. 373.

Figs. 2, 2a.—*Cuspidaria approximata*, E. A. Smith, Annals and Magazine of Natural History, (6) XVIII, 1896, p. 373.

Figs. 3, 3a.—*Myonera bicarinata*, E. A. Smith, Annals and Magazine of Natural History, (6) XVIII, 1896, p. 374.

Figs. 4, 4a.—*Scrobicularia ceylonica*, E. A. Smith, Annals and Magazine of Natural History, (6) XVIII, 1896, p. 375.

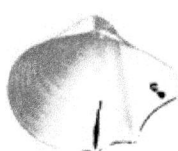

www.ingramcontent.com/pod-product-compliance
Lightning Source LLC
Chambersburg PA
CBHW022025190326
41519CB00010B/1603